Maharishi Dayanand Saraswati Book Series- 18

Creation Hymns of the Vedas

Prof. Ravi Prakash Arya

Chair Professor
Maharshi Dayanand Saraswati Chair (UGC)
Maharshi Dayanand University, Haryana

Amazon Books, USA

In Assiciation with

Indian Foundation for Vedic Science

H.O.1051, Sector-1, Rohtak, Haryana, India
Ph. 09313033917; 9650183260
Email: vedicscience@rediffmail.com; vedicscience@gmail.com
Web: https://vedic-sciences.com

First Edition
Kali era 5125 (Christian era 2023)
Kalpa era 1,97,29,49,125
Brahma era 15,55,21,97,29,49,125

ISBN No. 978-93-94724-11-2

© Author

Printed by

Indian Foundation for Vedic Science, 1051, Sector-1, Rohtak-
124001, Haryana

Contents

Introduction

The Vedas

The Vedas are the blueprint of the creation of the living and non-living world. This blueprint exists in Brahman and is manifested in the form of creation. It was perceived directly by the high-spirited yogī-s coded in the form of the Vedas. The process of creation of living and non-living beings is known as the first operation/Yajña, whereas the documentation of the blueprint of this Yajña of creation is called the second operation/Yajña. When the first operation/Yajña (creation) was over, the Devas (High-spirited yogī-s) performed the second operation/Yajña (documentation of the laws governing the creation of the living and non-living world). This fact is confirmed by the Vedas as follows:

'The Divine beings like Brahmā and other seers visualized the blueprint of creation (Yajña) that physically existed before them. This visualization was the first Dharma (unified law of creation). These laws of creation speak highly of the space of Brahman (Chidākāśa) where reside the divine beings and Siddhas aduring mokṣa[1].'

---

1 यज्ञेन यज्ञमयजन्त देवाः तानि धर्माणि प्रथमान्यासन् ।
ते ह नाकं महिमानः सचन्त यत्र पूर्वे साध्या सन्ति देवाः ॥

yajñena yajñam-ayajanta
devās-tāni dharmāṇi-prathamānyāsana
te ha nākam mahimānaḥ sacanta

In fact, all laws of creation visualized by the seers were coded in the forms of literary couplets or Chhandas. Those couplets or Chhandas were regarded as Dharmas, the laws governing the creation. This is why, Yāska (Nirukta, 1.20), an ancient Indian Vedic scholar alludes to the Origin of the Vedas (Chhandas) as:

'There were ṛṣis to whom was revealed Dharma (unified law governing creation, coded in the form of the Veda Mantras or Chhandas). They taught this dharma to their followers who were devoid of it, through oral instructions. When the later generation of ṛṣis was not able to retain this knowledge through oral instructions, all these laws were subjected to documentation just to aid them[2].

The records of the Vedas also tell us that on the basis of the first operation/yajña (creation) which occurred on account of oblations of all kinds of particles or forces, the origin of ṛcas contained in the Ṛgveda, sāmans contained in the Sāmaveda, yajuṣas contained in the Yajurveda and other Chhandas contained in the Atharvaveda took place in the cosmos. One of the seers of the Vājasaneyī Saṁhitā (31.7) sheds ample good light on this fact as:

'From the Yajña (creation), in which oblations of all particles and forces were made, originated Ṛcas, Sāmans,

---

yatra pūrve sādhyāḥ santi devāḥ.
(The Ṛgveda, 1.164.50; 10.90.16; the Atharvaveda, 7.5.1; the Vājasaneyī Saṁhitā, 31.16; the Taittirīya Saṁhitā, 3.5.11.5; the Taittirīya Āraṇyaka, 3.2.7)

2 साक्षात्-कृत्-धर्माणः ऋषयो: बभूवु:
sākṣāt-kṛt-dharmāṇaḥ ṛṣayoḥ babhūvuḥ.

Yajuṣas and other Chhandas (the Atharvaveda)[3].'

In fact, in that cosmic Yajña of creation, no material oblation was offered, as usual. However, ṛcas acted as the oblations of milk. Similarly, yajuṣas acted as oblations of Ghee, and the sāmans acted as the oblations of the Soma. This fact has been very carefully disclosed in the *Śatapatha Brāhmaṇa* (11.5.6.3,4,5). Here it may be pointed out that the sage to whom the knowledge of creation was revealed was Brahmā who passed this knowledge on to four ṛṣis, viz. Agni, Vāyu, Āditya, and Aṅgirā.

Classification of literary Couplets, or Chhandas into *Ṛk*, *Yajuḥ*, and *Sāman*:

As a result of the second great operation/Yajña (documentation of revealed knowledge in coded form) by the enlightened ṛṣi Brahmā, a huge number of couplets/Chhandas came into being dealing with Prakṛti or mass-energy existing at various levels of creation, viz. Bhūmi (Bhūtākāśa/visible space and planets), Dyau (Chidākāśa and stars), and field or Intervening space between Bhūmi and Dyau or planets and stars. For example,

1. Agni as geothermal energy in the earth and as mass-energy of Bhūtākāśa (visible space).

2. Field energy or Vāyu in the space intervening Bhūtākāśa and Chidākāśa, as well as planets and the stars.

3. Āditya or Sūrya as the light of stars and the inactive energy of the light space.

---

3   तस्माद् यज्ञात्सर्वहुतः ऋचः सामानि जज्ञिरे ।
छन्दांसि जज्ञिरे तस्माद् यजुस्तस्मादजायत ।।
tasmad yajñātsarvahutaḥ ṛcaḥ sāmāni jajñire.
chandāmsi jajñire tasmad yajus-tasmadajāyata.

Thus, the knowledge received about the geothermal energy of earth and mass-energy of Bhūtākāśa called (Agni) and its sub-forms was coded as ṛcās. The knowledge received about field energy (Vāyu), and electric force (Indra) and their forms existing in the space intervening planets and stars as well as Chidākāśa and Bhūtākāśa was coded as yajuṣas and the knowledge of the and solar energy, its sub-forms and inactive energy of Chidākaśa was coded as sāmans. The *Śvetāśvatara Upaniṣads* (6.18), the *Aitareya Brāhmaṇa* (25.7) and the *Manusmṛti* (1.23) had it as:

'To make the second great operation/Yajña a success, three types of (brahmas) Vedic couplets were derived. From Agni (geothermal energy of the planets and mass-energy of Bhūtākāśa) were derived ṛcas; from Vāyu (field energy of space intervening planets and stars, or the Bhūtākāśa and the Chidākāśa) were derived yajuṣas and from Sūrya (light of stars or inactive energy of Chidākāśa) were derived sāmans[4].

N.B.: Here Brahma means 'Veda' that is why a Brahmachārī is always known as one who undertakes the study of the Veda.

According to the Śatapatha Brāhmaṇa (11.5.8.3)

'On account of three forms of energy (tapta) in three spaces, three Vedas came into being. On account of Agni (geothermal energy of the planets and mass-energy of Bhūtākāśa) came into being; couplets called ṛcas of the *Ṛgveda*, on account of Vāyu (field energy or electric force of space intervening planets and stars; Bhūtākāśa

---

4   अग्नि वायुरविभ्यस्तु त्रयम् ब्रह्म सनातनम्
    दुदोह यज्ञ सिध्यर्थं ऋग्-यजुः-सामलक्षणम्
    agni vāyu-ravibhyastu trayam brahma sanātanam
    dudoha yajña siddyarathaṁ ṛg-yajuḥ-sāma-lakṣaṇam

and Chidākāsa) came into being; yajusas of the *Yajurveda* and on account of Sūrya (solar energy or inactive mass-energy of Chidākāsa) came into being the *Sāmaveda* or samans.'

The emergence of Saṁhitā literature

Later on, couplets dealing with ṛcas, yajusas, sāmans, and other chhandas were consolidated and compiled ṛṣi-wise and deity-wise into different literary styles and classified as the Ṛk Saṁhitā, the Yājuṣa-Saṁhitā, and the Sāma Saṁhitā respectively regardless of their earlier classification. During the course of the new classification, knowledge coded in metrical style was compiled as the *Ṛk Saṁhitā* (*ṛgarcani*). Knowledge coded in prose style was separately compiled in the name of the *Yājuṣa Saṁhitā* (*yat praśliṣṭaṁ paṭhitaṁ tat yajuḥ*) and the knowledge coded to the tune of cosmic vibrations was compiled under the caption the *Sāma Saṁhitā*. (*gitiṣu sāmākhyā*). The knowledge of miscellaneous nature related to the life of human beings coded metrically was compiled under the title of *Atharva Saṁhitā*. So, the *Atharva Saṁhitā* became the representative of all other three Saṁhitās.

Creation

The creation is the karma (action) of Brahman. Brahman is also called Kavi in the Upaniṣads. It is said that: कवेर्कर्म काव्यम्, i.e. Karma of Kavi is called Kāvya. So, this creation is the Kāvya of Brahman. Veda is the vāk (expression of the karma of the Brahman). Since this creation is described by Veda, so Veda is called Vāk of Brahman. Thus, the karma of Brahman exists in two forms:

1. In the visible form as a universe.

2. In the vāk form as Veda.

Everybody sees his karma in the form of creation, and hears his karma in the form of vāk (Veda), but very few except Brahmā are capable of comprehending it. Everybody is not able to hear this vāk. That is why in the *Ṛgveda* (10.71.4), it is said:

उत त्वः पश्यन्न ददर्श वाचमुत त्वः श्रृण्वन्न श्रृणोत्येनाम् ।
उतो त्वस्मै तन्वंश् वि सस्ने जायेव पत्य उशती सुवासाः ॥

uta tvaḥ paśyanna dadarśa vācham
uta tvaḥ śṛṇvanna śṛṇotyenām
uto tvasmai tanvaṁ visasre
jāyeva patya uśatī suvāsā

There are some who on seeing his karma in the form of creation do not see (understand) it, while another hearing his karma in the form of Veda vāk does not hear (understand it). His action in the form of creation and in the form of Veda vāk unfolds itself to the deserving seer, like the wife to her husband.

In other words, it can be said that act of creation of Brahman is expressed by His vāk (Veda) and the vāk of Brahman is the blueprint of creation in his Him. Thus, both are mutually interdependant. The 'Ṛta' (the law of dynamism) remains the same at all levels of nature and the universe, while their expression as we know is quite varied. The underlying laws of Ṛta applied to Satya (eternal prakṛti) are the same even when they express themselves in different forms, shapes, components, and levels of sophistication, complexity, and completeness. The Veda has expressly disclosed this fact, as:

एकं सद् विप्रा बहुधा वदन्ति ।

ekam sad viprā bhudhā vadanti

There is only one underlying law or truth expressed in different forms due to the law of dynamism.

In its most abstract form, the law of Ṛta (dynamism) is nothing but the 'will of Brahman' that moves or activates the 'Satya' (prakṛti or inactive energy) to undergo vikṛti active state. This 'Will of Brahman', in its true sense, can also be appreciated through mathematical formulae, laws of physics, chemistry, biology, anatomy, or physiology, but can also be understood in terms of family, social and national structures. The manifestation of unified law in various forms takes place due to the factor of māyā.

इन्द्रो मायाभिः पुरुरूप ईयते ।

indro māyābhiḥ pururūpa īyate.

If one is able to transcend this māyā factor in the universe, known variously as the 'unified law' or the 'will of Brahman' or the 'Veda', it will unfold itself to him. This unified law is the subject of subjective experience, and its manifestation into various forms of the universe is the subject of objective assessment. It exists in the objectively assessed physical universe as a 'unified field', but on the subjective level of experience, it unfolds itself as Brahman (pure consciousness, self-referral soul (pure Being), and ultimate truth. So, the unified field and unified law are the two sides of the same coin. This creation is a unified field and Veda is the unified law. Or say that in the context of subjective internal experience, it is the 'unified law' called Veda, but in the context of the objective external universe, it is a 'unified field' in its abstract form.

Vedic visionaries/seers who were the masters of vision (*ṛṣyo mantradraṣṭāraḥ*) visualized the unified law up to the minutest details beyond the limits of time and space in Samādhi. That unified law is called Veda. The mysteries of nature and the universe are unraveled in the Vedic mantras (the direct or primary knowledge acquired by the seers during Samādhi). Not only the Vedic ṛṣis visualized the unified law underlying the creation, they also picturized the state of being before creation.

After a brief introduction to the Vedic model of cosmology, the various creation sūktas of the Vedas depicting the condition before creation, the process of creation are given here with their scientific translations. I hope this scientific translation of the creation sūktas will help scholars, researchers, theoretician physicists and readers and have an actual and factual estimate of the Vedic theory of creation.

Puruṣa Sūkta

(Ṛgveda 10.90; Yajurveda 31.1-16; Atharvaveda 19.6)

In Puruṣa Sūkta, the word Puruṣa is used in the Vedas in three meanings.

1. It is the name of an individual embodied soul.

2. It is the name of Brahman and also the name of Chidākāśa where mass-energy of the Universe is located in inactive form. When Brahman takes saṅkalpa of creation, prakṛti (inactive energy) gets agitated and one fourth of inactive energy/dark energy converts into active energy and participate in the creation. The space of active energy is called Ākāśa or Bhūtākāśa. Here it may be understood that Chidākāśa is Avakāśa appearing as vacuous space where energy is located in inactive or dark form called 'adṛṣṭa' (invisible) and Bhūtākāśā is space with active energy which is dṛṣṭa (visible) and forms the dṛśya jagat (visible world). The same is also called 'Brahmaṇḍa Puruṣa'.

3. It is also used to denote the total mass-energy of chidākāśā (inactive or dark energy) and Bhūtākāśa (active energy).

सहस्रशीर्षा पुरुषः सहस्राक्षः सहस्रपात् ।
स भूमिꣳसर्वत स्पृत्वाऽत्यतिष्ठद्दशाङ्गुलम् ॥ य॰ 31.1

sahasraśīrṣā puruṣaḥ sahasrākṣaḥ sahasrapāt,
sa bhūmiꣳsarvata spṛtvā'tyatiṣṭhaddaśāṅgulam.

Spiritual Meaning: Puruṣa (embodied soul), has taken infinite of births since the beginning of creation, thereby manifesting in infinite number of heads, infinite number of eyes, and infinite number of feet; it has taken over in its possession this material body of 10 organs (five sense and five motor) through his birth.

Scientific Meaning: [Puruṣaḥ] Brahman or Chidākāśa is [atyatiṣṭhat] situated beyond [daśāṅgulam bhūmiṁ] the bhūtākāśa made up of five sukṣma bhūtās (tanmātrās) and five mahābhūtas (gross evolutes of prakṛti) or 10 dimensional bhūtākāśa [spṛtvā] by covering it [sarvataḥ] from all sides. [Saḥ] He has pervaded this bhūtākāśa containing [sahasra-śīrṣā] infinite number of heads of living beings and infinite number of galaxies in the material world; [sahasrākṣaḥ] infinite eyes so far as living beings are concerned and infinite number of stars so far as this material world is concerned; [sahasra-pāt] infinite number of feet so far as living beings are concerned and infinite numbers of planets so far as this material world is concerned.

Note: In Vedic Science, as pointed out above, there are two types of spaces: chidākāśa (space of Brahman containing dark energy or inactive energy) and bhūtākāśa (containing active energy). Bhūtākāśa, often called Brahmāṇḍa, is covered from all sides by chidākāśa. Bhūtākāśa is ten dimensional and made up of five sukṣma bhūtas (tanmātras, i.e. śabda, sparṣa, rūpa, rasa, and gandha) and five mahābhūtas (gross evolutes, i.e. visible ākāśa, vāyu, agni, jala and pṛthivī.

पुरुष एवेदं यद्भूतं यच्च भाव्यम् ।
सर्वं उतामृतत्वस्येशानो यदन्नेनातिरोहति ॥ यजु॰ 32.2

puruṣa ēvedaṁ sarvaṁ yadbhūtaṁ yachcha bhāvyam,
utāmṛtatvasyeśāno yadannenātirohati.

2. **Spiritual Meaning:** Puruṣa (soul) is the source of present living bodies, whatever has born and whatever

yet to be born; it is attains immortality by transcending the material body.

Scientific Meaning: [Puruṣa] Mass-energy of universe is the source of [yat idam] this present visible world [yat bhūtam] whatever has been created and [yat bhāvyam] whatever yet to be created; it is also [īśānaḥ] owner [amṛtatvasya] of inactive energy which [atirohati] leaves its actual inactive form and assumes the active form [annena] for the sake of creating the visible world which is the means of bhoga [enjoyment] for the living beings.

एतावानस्य महिमातो ज्यायांश्च पुरुषः ।
पादोऽस्य विश्वा भूतानि त्रिपादस्यामृतं दिवि ॥ यजु०31.3

etāvānasya mahimāto jyāyāṁścha puruṣaḥ,
pādo'sya viśvā bhūtāni tripādasyāmṛtaṁ divi.

Spiritual Meaning: [Asya] This body has a [etāvān] great [mahima] significance for the Puruṣa, as the mokṣa can only be attained with this body; and Puruṣa is [cha] even [jyāyān] superior [ataḥ] than this body; only the fourth part of its life-span, i.e. Gṛhastha Āśrama is meant for enjoying this [viśvā] whole [bhūtāni] material world and [tripāda] three parts [Brahmacharya, Vānaprastha and Sannyāsa] [asya] of its life are meant for [amṛtam] attaining immortality.

Scientific meaning: [etāvān] This Bhūtākāśa, the vast expanse of active energy tells [mahima] the vastness of [asya] this Chidākāśa, abode of inactive or dark energy. In fact, [puruṣaḥ] Chidākāśa is [api] even [jyāyān] greater [ataḥ mahimnaḥ] than the Bhutākāśa, the vast expanse of baryonic energy. [pādaḥ] One fourth [asya] of this energy [viśvā bhūtāni] takes part in the creation of universe; [tripāt] three fourth [asya] of it remains inactive or in dark form [divi] in dyauloka or chidākāśa.

Note: The mantra says that 25% of the energy

participate in creation, 75% energy is inactive or in dark form. During dissolution, energy becomes inactive, converting into dark energy. When energy is agitated by the saṅkalpa of Brahman, one fourth of it becomes active or baryonic energy. According to Vedic scientists, space with inactive energy is called chidākāśā/Avakāśa or vacuous space and space with active energy is called Ākāśa/Bhūtākāśa or space.

त्रिपादूर्ध्व उदैत्पुरुषः पादोऽस्येहाभवत्पुनः।
ततो विष्वङ् व्यक्रामत् साशनानशने अभि ॥ यजु० 31.4

tripādūrdhva udaitpuruṣaḥ pādo'syehābhavatpunaḥ,
tato viṣvan vyakrāmat sāśanānaśane abhi.

Spiritual Meaning: With the help of these three Āśramas of Brahmacharya, Vānaprastha and Saṁnyāsa, it (the soul) rises to the highest place (of immortality), but its excessive involvement in the other fourth part [Gṛhastha Āśrama] makes it cycle in the animate and inanimate world (the mundane world) repeatedly.

Scientific meaning: [Tripāt] Three-fourth of the [puruṣaḥ] energy [udait] is located [ūrdhvam] above the space or bhūtākāśa in inactive form in the chidākāśa. [pādaḥ] One fourth [asya] of this energy [abhavat] participate in creation [iha] in this universe [punaḥ] again and again, [tataḥ] and so [abhivyakrāmat] transforms [viṣvan] in diverse forms of [sāśane] animate and [anaśane] inanimate things in bhūtākāśa.

Note: The mantra says that only one fourth of the energy participate in creation. The three fourth of it remains inactive in the chidākāśa.

तस्मादविराळजायत विराजो अधि पुरुषः।
स जातो अत्यरिच्यत पश्चादभूमिमथो पुरः ॥ यजु० 31.5

tato virāḍajāyata virājo adhi puruṣaḥ,

sa jāto atyarichyata paśchādabhūmimatho puraḥ.

Spiritual Meaning: This soul [Puruṣa] due to its sanskāras of the visible world [ajāyata] occupies [virāṭa] the corporeal frame which is governed by it; the soul, before its birth was separate from the visible world, but soon after it's occupying the corporeal frame it considers it the part of visible world.

Note: The material body is called virāṭ.

Scientific meaning: [Tasmāt] From this Bhūtākāśa [ajāyat] was born [virāṭ] universe. [Puruṣaḥ] The Chidākāśa [adhi] transcends [virājaḥ] this universe. [Saḥ] The universe [jātaḥ] after its creation [atyarichyat] assumed its distinct form. [Paśchād] After the origin of bhūtākāśa [bhūmim] the stars and planets were formed [atha] followed by jīvas [puraḥ] attaining their corporeal frames in the universe.

Note: The Mantra says that Bhūtākāśa is the part of Chidākāśa, active energy provides it the distinct form from Chidākāśa.

तस्माद्यज्ञात्सर्वहुतः संभृतं पृषदाज्यम् ।
पशूँस्ताँश्चक्रे वायव्यानारण्या ग्राम्याश्च ये ॥ यजु॰ 31.6

tasmādyajñātsarvahutaḥ sambhṛtaṁ pṛṣadājyam,
paśūṁstāṁśchakre vāyavyānāraṇyā grāmyāścha ye.

Spiritual sense: The yajña of the bondage with the visible world, in which all souls were sacrificed, produced means of their enjoyments. It also made them to various types of creatures like birds flying in air, wild animals, plants and trees, and those that are grāmya (humans and tame animals).

Scientific meaning: [Tasmāt yajñāt] From the mental yajña of Brahman, i.e., sankalpa of Brahman [sarvahutaḥ], wherein inactive energy/dark energy was sacrificed into active energy/baryonic energy, [pṛṣat] curd, i.e. matter particles and [ājya] butter, i.e. antiparticles [sambhṛtam] were formed. Three types of [paśun] particles [chakre] were fromed [tān] which were [vāyavyān] field particles, [āraṇyāḥ] elementary particles, and [gramyāḥ] composite particles.

Note: According to the above mantra, there are three types of paśus or particles.

1. Vāvya paśus or field particles that live in field or intervening space. They are the transition state between conversion of dark energy/inactive energy into active or baryonic energy.

2. Āraṇya paśus or elementary particles are those that live alone.

3. Grāmya paśus or composite particles living in the company of others.

तस्माद्यज्ञात्सर्वहुत ऋचः सामानि जज्ञिरे ।
छन्दाꣳसि जज्ञिरे तस्माद्यजुस्तस्मादजायत ॥ यजु॰ 31.7

tasmādyajñātsarvahuta ṛchaḥ sāmāni jajñire,
chhandāꣳsi jajñire tasmādyajustasmādajāyata.

Spiritual meaning: This bondage with the visible world gave the souls knowledge of Ṛcās (geothermal energy of the planets), Sāmans (energy of the photosphere of the stars), Yajus (field energy or magnetosphere of the planets) and Atharvaveda (rules for living beings).

Note: Here the mantra says that the laws in society should be based upon natu ral laws.

Scientific meaning: [Tasmāt yajñāt] From the mental yajña of Brahman, i.e. saṅkalpa of Brahman [sarvahutaḥ], wherein oblation of inactive/dark energy was given, i.e. wherein inactive energy converted into active energy, [ṛchaḥ] the ṛcās (geothermal energy of planets), [sāmāni] the sāmans (energy of photosphere of the stars) was produced. [tasmāt] From that yajña [yajuḥ] field energy or magnetosphere of the planets and stars and ora or living beings [ajāyat] were formed and [tasmāt] from that yajña [chhandansi] the rest of Chhandas and other forms of energy were formed.

तस्मादश्वा अजायन्त ये के चोभयादतः ।
गावो ह जज्ञिरे तस्मात्तस्माज्जाता अजावयः ॥ यजु॰ 31.8

tasmādaśvā ajāyanta ye ke chobhayādataḥ,
gāvo ha jajñire tasmāttasmājjātā ajāvayaḥ.

Spiritual Meaning: This bondage made the souls born as endowed with [āśva] kṣatriya temperament, qualities and karmas, [gāvaḥ] Brāhamaṇa temperament, qualities and karmas, Vaiśya temperament, qualities, and karmas and Śudra temperament, qualities, and karmas; all of them are known as [ubhayādataḥ] dvijas.

Scientific meaning: [Tasmāt] It is from that yajña, i.e. from the Saṅkalps of Brahman [aśvāḥ] solar energy [ajāyanta] was formed [ye ke cha] which [ubhayādataḥ] has wave-particle duality; [tasmāt] from that yajña, [gāvaḥ] geothermal energy particles were [jañjīre] created; [tasmāt] from the same yajña, [ajā] the dark energy in Bhūtākāśa and [avi] field energy particles [jātāḥ] were formed.

Note: The mantra tells that owing to the Saṅkalpa of Brahman the inactive/dark energy of Chidākāśa was converted in active/baryonic energy of Bhūtākāśa forming various forms of it in Bhūtākāśa, like solar, geothermal, dark, and field energies.

तं यज्ञं बर्हिषि प्रौक्षन् पुरुषं जातमग्रतः ।
तेन देवा अयजन्त साध्या ऋषयश्च ये ॥ यजु॰ 31.9

taṁ yajñaṁ barhiṣi praukṣan puruṣaṁ jātamagrataḥ,
tena devā ayajanta sādhyā ṛṣayaścha ye.

Spiritual Meaning: That soul, which existed prior to occupying its corporeal frame, was sprinkled with barhis (sanskāras of the visible world), With the help of this material frame (physical body), enlightened scholars, yogīs and seers performed samādhi yajñā.

Scientific meaning: [Tam] Prakṛti or inactive energy of chidākaśa which was [yajñam] the ultimate source of creation yajña and [agrataḥ jātam] existed prior to this creation [praukṣan] and was agitated [barhiṣi] under power of the saṅkalpa of Brahman and consequently converted into active energy. [Tena] Due to this reason, [devaḥ] enlightened persons and [sādhyāḥ] siddhas (experts) and [ṛṣis] seers [ayajanta] worshiped [puruṣam] Brahman.

Note: The Mantra wants to say that energy was inactive/dark during the period of dissolution. Saṅkalpa of Brahman agitated it. Due to agitation by the saṅkalpa of Brahman, it was converted into active or baryonic form.

यत्पुरुषं व्यदधुः कतिधा व्यकल्पयन् ।
मुखं किमस्यासील्किं बाहू किमूरू पादा उच्येते ॥ यजु॰ 31.10

yatpuruṣaṁ vyadadhuḥ katidhā vyakalpayan;
mukhaṁ kimasyāsītkiṁ bāhū kimūrū pādā uchyete.

Spiritual Meaning: When the soul came under the bondage of the visible world and manifested in a material frame, into how many portions it was divided? What was the mouth of embodied soul called, what were arms, what his thighs and what were his feet called?

Scientific Meaning: [yat] When [puruṣam] Chidākāśa [vyadadhuḥ] was transformed into Bhūtākāśa by the saṅkalpa of Brahman, [katidhā], into how many portions the Bhūtākāśa [vyakalpayan] was divided? [Kim] What was [asya] its [mukham] mouth called, [kau] what were its [bāhū] arms, [ūrū] thighs, and [pādā] feet [uchyate] called?

ब्रह्मणोऽस्य मुखमासीद् बाहू राजन्यः कृतः ।
उरू तदस्य यद्वैश्यः पद्भ्याꣳ शूद्रो अजायत ॥ यजु॰ 31.11

brahmaṇo'sya mukhamāsīd bāhū rājanyaḥ kṛtaḥ;
urū tadasya yadvaiśyaḥ padbhyāꣳ śūdro ajāyata.

Spiritual Meaning: The mouth of the embodied soul was called Brāhmaṇa, its arms were made the Rājanya, its thighs became its Vaiśya part; it became the Śūdra from its feet or its feet formed its Śūdra part.

Note: The body of embodied soul was divided into four parts. From mouth, it was called Brāhamaṇa, from arms it became Kṣatriya, from thighs Vaiśya and from feet Śūdra. So, all the four varṇas are in one body of an embodied soul. Afterward, the social stratification was done into four varṇas on the basis of nature (temperament), quality and karmas of the people. Varṇa system was not based upon birth, but upon temperament, quality, and karma of a person. Caste system was based upon birth. Vedas and Vedic literature mentions the Brāhmaṇa Varṇa and not Brāhamaṇa caste.

Scientific Meaning: [Mukham] Fire (Light and heat of stars) [āsīt] was [brahmaṇaḥ] Brāhmaṇa part [asya] of Bhūtākāśa, [bāhū] directions [kṛtaḥ] were made [rājanyaḥ] its kṣatriya part, [ūrū] the midspaces acted as [tat asya] its [vaiśya] Vaiśya part; and [padbhyām] from the point of planets it [ajāyat] became śūdra or say planets were its Śūdra part.

Note: Constellations are the Brāhmaṇa part of Bhūtākāśa, directions are Kṣatrīya part, Midspaces are the

Vaiśya part and planets are Śudra part of the Bhūtākāśa.

चन्द्रमा मनसो जातश्चक्षो: सूर्य्यो अजायत ।

श्रोत्राद्वायुश्च प्राणश्च मुखादग्निरजायत ॥ यजु॰ 31.12

chandramā manaso jātaśchakṣoḥ sūryyo ajāyata,
śrotrādvāyuścha prāṇaścha mukhādagnirajāyata.

Spiritual Meaning: The mind of embodied soul was represented by the moon (satellite); its eyes were represented by the sun; its mouth bears the representation of Indra and Agni, and its breath was represented by vāyu (air) of the visible world.

Scientific Meaning: [Chandramā] The moon [jātaḥ] acted [manasaḥ] the mind of Bhūtākāśa; [sūryaḥ] the sun [ajāyat] acted [chakṣo] its eyes; [agnir] agni (light and heat) acted its [mukhāt] mouth; [vāyuḥ] the air [cha] and [prāṇaḥ] breath acted its [śrotrāt] ears.

नाभ्या आसीदन्तरिक्ष ᳵ शीष्र्णो द्यौ: समवर्त्तत ।

पदभ्यां भूमिर्दिश: श्रोत्रात्तथा लोकाँ२ अकल्पयन् ॥ यजु॰ 31.13

nābhyā āsīdantarikṣaᳵ śīrṣṇo dyauḥ samavarttata,
padabhyāṁ bhūmirdiśaḥ śrotrāttathā lokāṁ2
akalpayan.

Spiritual Meaning: Its navel was represented by mid-sphere (intervening space), its head by celestial sphere (Sun), feet by the earth, ear by the quarters of space, so its body (microcosm) is the representation of the universe (macrocosm). It is often said in Indian philosophical tradition: yad anṇḍe tad brahmāṇḍe. The body of the soul is a representation of the cosmos.

Scientific Meaning: [antarikṣam] The mid-sphere or intervening space between planets and stars [āsīt] formed [nābhyāḥ] its naval part; [dyauḥ] the solar sphere [samvartat] became [śīrṣṇaḥ] its head; [bhūmiḥ] the earth (planetary sphere) became [padbhyām] its feet; [diśaḥ] directions (dimension of space) formed its [śrotrāt] ears; [tathā] In this way it [akalpayan] formed [lokān] other worlds.

यत्पुरुषेण हविषा देवा यज्ञमतन्वत।
वसन्तोऽस्यासीदाज्यं ग्रीष्म इध्मः शरद्धविः॥ यजु॰ 31.14

yatpuruṣeṇa haviṣā devā yajñamatanvata;
vasanto'syāsīdājyaṁ grīṣma idhmaḥ śaradhaviḥ.

Spiritual Meaning: The mind and senses offered the oblation for this soul to expand its bondage with the mundane world. The Spring season (the childhood of the embodied soul) acted as the offering of ghī, Summer season (young age) acted as samidhā and the Autumn (old age) acted as the oblation of the yajña of its bondage with the visible world.

Scientific meaning: [yat] When [haviṣā] the oblations of [puruṣeṇa] dark/inactive energy were offered [devāḥ] baryonic energy particles are formed and [atanvata] the expansion of [yajñam] creation takes place. [ājyam] Inactive energy is [vasantaḥ] the Spring season [asya] of this yajña of creation; [idhmaḥ] samidhās are [grīṣma] the Summer season; and the [śarad] Autumn season is [haviḥ] the oblations of this yajña of creation.

सप्तास्यासन् परिधयस्त्रिः सप्त समिधः कृताः।
देवा यद्यज्ञं तन्वाना अबध्नन् पुरुषं पशुम्॥ यजु॰ 31.15

saptāsyāsan paridhayastriḥ sapta samidhaḥ kṛtāḥ;
devā yadyajñaṁ tanvānā abadhnan puruṣaṁ paśum.

Spiritual Meaning: Its body was made of seven dhatus, which are its paridhis (enclosures). Seven dhatus and three doṣas make 3×7= 21 samidhās for the yajña (function of its body). The mind and senses, while carrying out the yajña of interaction with the visible world, sacrificed the soul on the altar of this yajña.

Scientific meaning 1: [sapta] The seven lokas (Bhūḥ[5], Bhuvḥ[6], Svaḥ[7], Mahaḥ[8], Janaḥ[9], Tapaḥ[10], Satyam[11]) are the [paridhayaḥ] seven enclosers [asya] of this yajña of creation. [triḥ sapta] Mind, 5 sense organs, 5 motor organs, 5 tanmātras and 5 gross bhūtas = these 21 were [kṛtāḥ] made [samidhaḥ] samidhās (fire-wood) in this yajña. [puruṣam], energy [paśum] particles [abadhnan] were sacrificed in [yajñam] yajña of creation [yad] which was [tanvānā] extended by the [devāḥ] matter particles.

Scientific meaning 2: The yajña of creation on the earth is carried by the solar radiation. The sun encloses the earth at seven places[12] which are its seven enclosures or chhandas; thrice seven samidhās (21 samidhas are made by three spaces-earth (pṛthivī), midspace [antarikṣa] and dyau (solar region) multiplied by seven

---

5 Bhūḥ is the Pṛthivī loka (the earth).

6 Bhuvaḥ is the space between the earth and the sun.

7 Svaḥ is the Sūrya loka or Solar sphere (the sun).

8 Mahaḥ is the Space between the Sun and our Galaxy.

9 Janaḥ is the Prajāpati loka (Our Galaxy).

10 Tapaḥ is the space between Prajāpati loka (Galaxy) and Svaṁbhūva loka (Super galactic centre).

11 Satyam loka is Svaṁbhūva loka (Super galactic centre)

8   a). 23.5⁰ South of Equator on 22 Dec.,

   b). 20⁰ South of Equator on 21 January,

   c). 12⁰ South of Equator on 18 Feb.

   d). 0⁰ or on Equator on 21 March,

   e). 12⁰ North of Equator on 20 April,

   f). 20⁰ North of Equator on 21 May,

   g). 23.5⁰ North of Equator on 21 June.

enclosures. For the smooth conduct of yajña of creation on the earth, Puruṣa (solar energy) is [badhnan] captured by the magnetosphere of earth.

यज्ञेन यज्ञमयजन्त देवास्तानि धर्माणि प्रथमान्यासन् ।
ते ह नाकं महिमानः सचन्त यत्र पूर्वे साध्याः सन्ति देवाः ॥ य० 31.16

yajñena yajñamayajanta devāstāni dharmāṇi
prathamānyāsan,
te ha nākaṁ mahimānaḥ sachanta yatra pūrve sādhyāḥ
santi devāḥ.

Spiritual Meaning: With this corporeal frame, the divine souls born first at the beginning of creation performed the yajña of attaining the knowledge of the creation of the visible world, this knowledge was called the first dharma (rules or principles underlying the process of creation, and duties of created beings) or Veda. Those first born great divine souls became partakers of the immortality where the ancient Yogīs and divine souls abide.

Scientific Meaning: [Yajñena] The yajña of transformation of inactive energy into active energy [ayajanta] resulted [yajñam] into creation of [devāḥ] baryonic energy particles. [tāni dharmāṇi prathamānyāsan] Those baryonic/active energy particles were the first material sustaining this creation. [te] Those [ha] active energy particles have their source [nākam] in the space of Brahman (chidākāśa) [yatra] where they [santi] are located [pūrve] in their prototype form (inactive/dark form) [sādhyāḥ] as source [devāḥ] of the active particles.

Here the author of the Nirukta, Yāska, says:

यज्ञेन   यज्ञमयजन्त   देवाः,   अग्निनाग्निमयजन्त   देवाः,   अग्निः
पशुरासीत्तमालभन्त तेनायजन्तेति च ब्राह्मणम् । तानि धर्माणि प्रथमान्यासन्, ते ह

नाकं महिमानः समसेवन्त, यत्र पूर्वे साध्याः सन्ति देवाः, साधनाः, द्युस्थानो देवगण इति नैरुक्ताः ॥ निरु. 12.41

yajñena yajñamayajanta devāḥ, agnināgnimayajanta devāḥ, agniḥ paśurāsīttamālabhanta tenāyajanteti cha brāhmaṇam / tāni dharmāṇi prathamānyāsan, te ha nākaṁ mahimānaḥ samasevanta, yatra pūrve sādhyāḥ santi devāḥ, sādhanāḥ, dyusthāno devagaṇa iti nairuktāḥ.

[Devāḥ] Baryonic/Active energy particles [yajñena] converting into matter particles [ayajanta] carried out [yajñam] the process of creation. In this process of yajña (creation) [devāḥ] the particles of 'agni' (baryonic energy) converted into matter particles. Energy was in the form of Paśus (particles). 'Those energy particles were sacrificed to be converted into matter particles; this is stated in the Brāhmaṇa texts. Those particles were the first [dharma] material sustaining this creation. They were located in the nāka (space of Brahman/chidākāśa), where did they resided in their prototype form/dark or inactive energy form as source. According to the school of Etymologists, Devas (both energy and enlightened ones) are 'dwellers of [dyusthāna] the space of Brahman.

अद्भ्यः संभृतः पृथिव्यै रसाच्च विश्वकर्मणः समवर्त्तताग्रे ।
तस्य त्वष्टा विदधद्रूपमेति तन्मर्त्त्यस्यदेवत्वमाजनमग्रे ॥ यजु० 31.17

adbhyaḥ sambhṛtaḥ pṛthivyai rasāchcha viśvakarmaṇaḥ samavarttatāgre, tasya tvaṣṭā vidadhadrūpameti tanmarttyasyadevatvamājanamagre.

[Agre] In the beginning, [pṛthivyaiḥ] for the formation of matter, [rasāt] due to the will-power [Viśvakarmaṇaḥ] of Creator, [adbhyaḥ] from active energy particles [samvarttat] matter particles were formed. [sambhṛtaḥ] This whole universe is sustained by those matter particles. After formation of matter, [tvaṣṭā] The Almighty designer [eti] keeps on [vidadh] designing [viśva-rūpam] various forms [tasya] of this universe. [tat] His [ājānam] realization [devatvam] is called divine hood [martasya] of a human being. That is, by realizing Brahman a human being attains divinity or mokṣa.

वेदाहमेतं पुरुषं महान्तमादित्यवर्णं तमसः परस्तात् ।
तमेव विदित्वाति मृत्युमेति नान्यः पन्था विद्यतेऽयनाय ॥ य० 31.18

vedāhametaṁ puruṣaṁ mahāntamādityavarṇaṁ
tamasaḥ parastāt,
tameva viditvāti mṛtyumeti nānyaḥ panthā
vidyate'yanāya.

[Aham] I [veda] realize this Creator Puruṣa or Brahman [mahāntam] who is exalted above all, [ādityavarṇam] glorious like the Sun and [tamasaḥ parastat] beyond the darkness of ignorance. [viditvā] By realizing [tameva] Him alone, [mṛtyum eti] is death conquered. Except this [vidyate] there is [na anyaḥ] no other [panthā] road [ayanāya] leading to Salvation.

This verse is the answer to the question, "By knowing what can you become enlightened?" The answer is- 'I am certainly enlightened because I have realized the Puruṣa (the Supreme Lord) whose attributes have been described above, Who is the greatest of all, the oldest, Self-effulgent, above and beyond the darkness of ignorance and material world.'

No one can become enlightened without realizing Him because, by realizing the Puruṣa (the Supreme Lord) alone, can a man cross death and attain to that state of the highest bliss (Mokṣa) which is beyond death. There is no other means of reaching that state.

प्रजापतिश्चरति गर्भे अन्तरजायमानो बहुधा वि जायते ।

तस्य योनिं परि पश्यन्ति धीरास्तस्मिन्ह तस्थुर्भुवनानि विश्वा ॥ यजु॰ 31.19

*prajāpatiścharati garbhe antarajāyamāno bahudhā vi
jāyate,*

*tasya yoniṁ pari paśyanti dhīrāstasminha
tasthurbhuvanāni viśvā.*

[Prajāpati] The Lord of creatures [charati] pervades
[antar garbhe] inside of all animate beings and inanimate
things. [Ajāyamānaḥ] Being Himself unborn, He makes
this universe [vijāyate] manifest [bahudhā] in motley
forms. [Dhirāḥ] The wise [pari-paśyanti] visualize [tasya]
His [yonim] true Nature. [Tasmin] In Him do [viśvā] all
[bhuvanāni] worlds [tasthuḥ] find their stay.

यो देवेभ्य आतपति यो देवानां पुरोहितः ।

पूर्वो यो देवेभ्यो जातो नमो रुचाय ब्राह्मये ॥ यजु॰ 31.20

yo devebhya ātapati yo devānāṁ purohitaḥ,

pūrvo yo devebhyo jāto namo ruchāya brāhmaye.

[Namaḥ] Obeisance to the [ruchāya] illumination
[Brāhmaye] of Brahman, [yaḥ] Who [ātapati] shines forth
[devebhyaḥ] for the enlightened persons; [yaḥ] Who
[purohitaḥ] is the Chief Benefactor [devānām] of the
enlightened persons and [yaḥ] Who [pūrvaḥ jātaḥ] pre-
existed [devebhyaḥ] of all enlightened persons.

रुचं ब्राह्मं जनयन्तो देवा अग्रे तदब्रुवन् ।

यस्त्वैवं ब्राह्मणो विद्यात्तस्य देवा असन्वशे ॥ यजु॰ 31.21

ruchaṁ brāhmaṁ janayanto devā agre tadabruvan,

yastvaivaṁ brāhmaṇo vidyāttasya devā asanvaśe.

[Devaḥ] Enlightened persons [janyantaḥ] having
acquired this [rucham] lovable [brāhman] divine
knowledge [abruvan] explained [tat] it [agre] further (to
others). [Devāḥ] The enlightened person, [yaḥ tu] who
[vidyāt] knows [evaṁ] it as it is, [asan] has [devāḥ] all his

senses [vaśe] under his control.

श्रीश्च ते लक्ष्मीश्च पत्न्यावहोरात्रे पार्श्वे नक्षत्राणि रूपमश्विनौ व्यात्तम् ।
इष्णन्निषाणामुं म इषाण सर्वलोकं म इषाण ॥ यजु॰ 31.22

*śrīścha te lakṣamīścha patnyāvahorātre pārśve
nakṣatrāṇi rūpamaśvinau vyāttam,
iṣṇanniṣāṇāmum ma ivāṇa sarvalokaṁ ma iṣāṇa.*

(O God) [śrī] grace and [lakṣmī] lordship or sovereignty are like [te] your [patnyau] two wives; [ahorātre] the day [creation] and night [dissolution] are your [pārśve] two sides or parts; [nakṣatrāṇi] the constellations are your [rūpam] beauty; and [aśvinau] the two Aśvins are your [vyāttam] open mouth (the visible faces); that is why the constellations are counted from Aśvins. [Iṣāṇa] May your will bless [me] me with [amum] that final beatitude and [sarvalokam iṣāṇa], bless with all true happiness here and [iṣāṇa] bless [me] me with [iṣṇan] whatever desired by me.

Nāsadiya Sūkta

(Ṛgveda 10.129)

The state of being before creation is described in Nāsadīya sūkta. Its Devatā is Bhāvavṛtta (narration of creation in abstract form), as before creation everything remains merged in Brahman.

नास॑दासीन्नो सदा॑सीत्तदानीं॑ नासी॒द्रजो॒ नो व्यो॑मा प॒रो यत् ।

किमाव॑रीवः॒ कुह॒ कस्य॒ शर्म॒न्नम्भः॒ किमा॑सी॒द्गहनं॑ गभी॒रम् ॥

nāsadāsīnno sadāsīttadānīṁ nāsīdrajo no vyomā paro yat,
kimāvarīvaḥ kuha kasya śarmannambhaḥ
kimāsīdgahanaṁ gabhīram.

1. [Tadānīm] Before creation [na āsīt] there was neither [asat] absence [na sat] nor presence of matter [rajaḥ] the motion [na āsīt] was not there; there was [na] no [vyoma[13]] Bhūtākāśa (space), no midspace between the earth and the sun or any and [paraḥ[14]] beyond [yat] that, i.e. inter stellar space, as they were not present then. [kim] How could there be [āvarīvaḥ] any layers/substratum of this universe? [kuhaḥ] And where [kasya] and who [śarman] enjoyed pain and pleasure? (Means like bhogya prakṛti, there was no bhoktā jīva during dissolution). [ambhaḥ] Active energy/vikṛti heterogeneous state of sattva, rajas and tamas did not

---

13. व्योम अन्तरिक्षं (सायण)

14. परः व्योम्नः परस्तादुपरिदेशे द्युलोकप्रभृतिसत्यलोकान्तं यत् अस्ति तदपि नासीद्। (सायण)

exist [gahanam] under these complex and [gabhiram] unfathomable conditions?

Note: According to Vedic science, vyoma is antariksa (midspace between the earth and the sun) or say magnetosphere of the earth). It is also Bhūtākāsa where mass energy is in active form. This universe has 14 layers, so called *Chaturdaśabhuvana Garbha Brahmāṇḍam.*

न मृत्युरासीदमृतं न तर्हि न रात्र्या अह्न आसीत्प्रकेतः ।
आनीदवातं स्वधया तदेकं तस्माद्धान्यन्न परः किं चनास ॥ २ ॥

na mṛtyurāsīdamṛtaṁ na tarhi na rātrayā ahna
āsītpraketaḥ,
ānīdavātaṁ svadhayā tadekaṁ tasmāddhānyanna
paraḥ kiṁ chanāsa.

2. [Na] There was neither [mṛtyuḥ] death [na] nor [amṛtam] immortality [tarhi] at that period; [na āsīt] there was neither [apraketaḥ] differentiation of [ahnaḥ] day (creation) and [rātryā] night or dissolution; [tadekam] That One, Brahman, was there [svadhayā] sustained by his own power [anādivātam] without breathing air; [kiṁ ca na āsa] there was nothing [paraḥ] beyond [tasmāt] Him.

1. Before creation, only Brahman (God) was present. That time only Brahmākāśa (Space of Brahman) existed, bhūtākāśa (space of modern physics) was merged with Brahmākāśa.

तम आसीत्तमसा गूव्हमग्रेऽप्रकेतं सलिलं सर्वमा इदम् ।
तुच्छ्येनाभ्वपिहितं यदासीत्तपसस्तन्महिनाजायतैकम् ॥3 ॥

tama āsīttamasā gūḷhamagre'praketaṁ salilaṁ sarvamā
idam,
tuchchhyenābhvapihitaṁ
yadāsīttapasastanmahinājāyataikam.

3. [Agre] Before the creation [āsīt] there was [tama] inactive energy or dark energy and absence of calculable time; everything was [gudham] wrapped [tamasā] in dark energy and in timeless time; [idam sarvam] this entire [salilam] rarefied prakṛti or dark energy was [apraketam] undifferentiated, i.e. there was no difference of sattva, rajas and tamas; [ābhu] all sides of chidākāśa or space of Brahman [apihitam] were filled [tucchyena] with rarefied inactive, i.e. dark energy. [yad] Whatever existed before creation in inactive form [tat] that [āsīt] was [ajāyat] activated [ekam] solely [mahinā] by power of [tapas] saṅkalpa of Brahman.

Note: The above mantra indicates:

1. There are two forms of time, kāla (calculable time) and akāla (incalculable time or timeless time). Kāla has been identified with light and akāla or has been indetified with darkness.

2. Similarly there are two forms of Energy. 1). Prakṛti (inactive energy or dark energy) which do not participate in creation 2. Vikṛti (active energy) participates in creation. Prakṛti state is known as natural state or state with zero entropy. In natural state, the energy is inactive/rarefied and evenly distributed. In prakṛti state, force of attraction, distraction is always absent. Space with inactive energy is called in Vedic philosophical terms is called 'Avakāśa', 'Chidākāśa', or Space of Brahman. Elementary particles are formed from the active energy. In Vaiṣeśika the inactive energy particle is called 'Paramāṇu'. In Sāṅkhya, it is called Prakṛti. According to Vedic Science, in Vikṛti state, the inactive energy transforms into active energy particles or quanta form. The active energy particles are governed by the force of

attraction and distraction etc. In Vikṛti state, entropy starts increasing.

3. Before creation, energy was inactive and in rarefied form evenly distributing in the Ākāśa (space). Ākāśa (space) with inactive energy or dark energy is called 'Avakāśa' or 'Chidākāśa' (space of Brahman). We can also say that before creation, Avakāśa or Chidākāśa had uniform density everywhere.

The saṅkalpa of Brahman was at work behind the manifestation of universe.

कामस्तदग्रे समंवर्तताधि मनंसो रेतं: प्रथमं यदासीत् ।
सुतो बन्धुमसंति निरंविन्दन्हृदि प्रुतीष्यां कुवयों मनीषा ॥4 ॥

kāmastadagre samavartatādhi manaso retaḥ
prathamaṁ yadāsīt,
sato bandhumasati niravindanhṛdi pratīṣyā kavayo
manīṣā.

4. [Agre] Before creation, [kamaḥ adhi] the saṅkalpa of creation [yad] that [samvartat] arose in Brahman, [tad] [āsīt] was [prathamam] the first [manasaḥ retaḥ] mental seed or sanskāra of creation; [kavayaḥ] yogīs who know the past and future [hṛdi] in their samādhi [buddhyā pratīṣya] through ṛtambhara prajñā [niravindat] found [satah] inactive energy [bandhum] converting [asati] into active energy.

Note: In the Vedic science, inactive energy or prakṛti is called sat, and vikṛti or active energy is called asat. The saṅkalpa of Brahman in terms of the sanskāra of Brahman causes the activation of energy.

Here the phrase '*sato bandhum asati*' also explains the creation of living beings. Accordingly, the meaning will be as under:

[satah] Present life of living souls is [bandhu] bound with [asati] past unmanifest karmic sanskāras.

Two things are evident from the above mantra:

1. Creation is caused by the saṅkalpa or the sanskāra of Brahman. When Brahman takes saṅkalpa of creation, the inactive energy is activated. Space with active energy is called Ākāśa or bhūtākāśa and space where energy is inactive is called Avakāśa, Chidākāśa.

2. Present life of embodied souls is based upon their past life unmanifest kārmika sanskāras (karmas that have not fructified so far). According to Vedic philosophy, sanskāras of past life act as seed (information) for the creation of present life of souls.

तिरश्चीनो वितंतो रश्मिरेषामध: स्विंदासी३दुपरिं स्विदासी३त् ।

रेतोधा आंसन्महिमानं आसन्स्वधा अवस्तात्प्रयंति: पुरस्तांत् ॥ 5 ॥

tiraśchīno vitato raśmireṣāmadhaḥ svidāsī3dupari svidāsī3t,
retodhā āsanmahimāna āsansvadhā avastātprayatiḥ parastāt.

5. [Raśmi] Just as the rays of the sun pervade the entire universe quickly when it rises, similarly when energy is activated [eṣām] its particles quickly [vitataḥ] pervade the entire Bhūtākāśa and stretch out everywhere [tiraśchinaḥ] in middle, [adhaḥ svidāsīt] below, or [upari svidāsīt] above. In this creation at one hand, there [āsan] are Jīvas [retodhā] who are doers of karmas that act as the seed for their next lives and who are also bhoktas, i.e. enjoyers of this material creation. On the other hand, there is [mahimānaḥ] expanse of matter which is bhogya-prapañcha (means of enjoyment for the Jīvas). Between the two [prayatiḥ] bhoktas, i.e. Jīvas and [svadhā] bhogya-prapañcha (expanse of matter), bhokta, i.e. Jīvas

are [parastāt] superior and [svadhā] bhogya-prapañcha is [avastāt] inferior.

Note: In the above mantra, it has been informed that the whole creation is divided into bhoktas/Jīvas (living beings) and bhogya-prapañcha (material expanse). Living beings are superior to material expanse.

In the Vedic Science, matter is called anna or soma (food) and Jīva is called annāda or agni (eater of food).

एतद्वा इदमन्नं चैवान्नादश्च सोम एवान्नमग्निरन्नाद (Ś.Br. 1.4.2.13)

ētadvā idamannam chaivānnādaścha soma ēvānnamagnirannāda

को अद्धा वेद क इह प्र वोचत्कुत आजाता कुतं इयं विसृष्टिः ।
अर्वाग्देवा अस्य विसर्जनेनाथा को वेद यतं आबभूवं ॥6॥

ko addhā veda ka iha pra vochatkuta ājātā kuta iyam visṛṣṭiḥ,
arvāgdevā asya visarjanenāthā ko veda yata ābabhūva.

6. [Kaḥ] Who [addhā] really [veda] knows? [Kaḥ] Who [pravochat] will tell [iha] here [kutaḥ] from what material cause [ājātā] this creation came into being; [kutaḥ] from what efficient cause [iyam] it [visṛṣṭi] was created? [Devāḥ] The divine seers were born [arvāg] after [asya] this [visarjanena] material creation; [atha] so [kaḥ] only Prajāpati Brahman [veda] knows [yat] whence [ābabhūva] it arose?

Note: In the above mantra, it is clarified that every effect has its material (upādāna or samvāyi-kāraṇa) as well as efficient (nimitta-kāraṇa or asamvāyī-kāraṇa) cause.

इयं विसृष्टिर्यत आबभूव यदि वा दधे यदि वा न ।
यो अस्याध्यक्षः परमे व्यौमन्सो अङ्ग वेद यदि वा न वेद ॥7 ॥

iyaṁ visṛṣṭiryata ābabhūva yadi vā dadhe yadi vā na,
yo asyādhyakṣaḥ parame vyomanso aṅga veda yadi vā
na veda.

7. [Yataḥ] From which efficient cause [iyam] this [visṛṣṭi] creation [ābabhūva] came into being, that efficient cause, i.e. Brahman [yadi vā dadhe] governs it, [yadi vā na] if He does not govern it, who else does it. [yaḥ] He, Who is [asya] its [adhyakṣaḥ] Governor, resides [parame vyoman] in the supreme space (Chidākāśa or space of Brahman); [saḥ] He [aṅga] assuredly knows it, [yadi vā] if he [na] doesn't [veda] know it, who else does?

Note: In this mantra, Brahman is described as the (nimmitta-kāraṇa) efficient cause of creation. Efficient cause can only know its creation. He is also described as living in His own space called Parama Vyoma or Space of Brahman or Chidākāśa. Chidākāśa has the nature of 'Avakāśa', i.e., vacuum and not the Ākāśa (space). Ākāśa (space) has active energy, but Avakāśa (vaccum) has inactive or dark energy.

Hiraṇyagarbha Sūkta

(Ṛgveda 10.121)

हिरण्यगर्भः समवर्त्तताग्रे भूतस्य जातः पतिरेक आसीत् ।
स दाधार पृथिवीं द्यामुतेमां कस्मै देवाय हविषा विधेम ॥ 1 ॥

ऋ. 10.121.1; यजु. 13.4

hiraṇyagarbhaḥ samavarttatāgre bhūtasya jātaḥ
patireka āsīt,
sa dādhāra pṛthivīṁ dyām utemāṁ kasmai devāya
haviṣā vidhema.

[Hiraṇyagarbha] God with potential energy of universe in His womb [samavartata] existed [agre] in the beginning [bhūtasya] of this material creation. He [āsīt] was the [eka] only [pati] Governor or efficient cause of it. [Saḥ] He [dadhāra] sustains [pṛthivīṁ] this Bhūtākāśa (space with active energy and matter) [uta] and [dyām] Chidākāśa or Avakāśa or vacuous space with inactive or dark energy and matter; do we offer oblations unto Him.

य आत्मदा बलदा यस्य विश्व उपासते प्रशिषं यस्य देवाः ।
यस्यच्छायाऽमृतं यस्य मृत्युः कस्मै देवाय हविषा विधेम ॥ 2 ॥

ऋ. 10.121.2; यजु. 25.13

ya ātmadā baladā yasya viśva upāsate praśiṣaṁ yasya
devāḥ.
yasyacchāyā'mṛtaṁ yasya mṛtyuḥ kasmai devāya
haviṣā vidhema.

[Yaḥ] He who [ātmadā] activates souls and [bala dā] prakṛti [inactive energy] for the creation of living and non-living world in the beginning of the creation. [viśve] Everything that is born in this universe [upāsate] follows [yasya] His [praśiṣaṁ] laws; [yasya] His [chhāyā] refuge [amṛtam] provides emancipation; and [yasya] His desertion [mṛtyuḥ] causes mortality; [kasmai] unto Him, [devāya] the Blissful One and All-knowing, [vidhema] do we offer [haviṣā] our oblations.

यः प्राणतो निमिषतो महित्वैक इद्राजा जगतो बभूव ।
य ईशे अस्य द्विपदश्चतुष्पदः कस्मै देवाय हविषा विधेम ॥

ऋ. 10.121.3; यजु. 23.3

yaḥ prāṇato nimiṣato mahitvaika idrājā jagato babhūva,

ya īśe asya dvipadaś catuṣpadaḥ kasmai devāya haviṣā vidhema.

[Yaḥ] He, [mahitvā] in His infinite capacity, is [eka it] single [rājā] Lord of [prāṇataḥ] living and [nimiṣataḥ] non-living [jagat] world; [yaḥ] He [īśe] rules [dvipadaḥ] bipeds (two footed human beings) and quadrupeds (four footed animal beings); [kasmai] unto Him, [devāya] the Blissful One, [vidhema] do we offer [haviṣā] our oblations from the core of our heart and soul.

यस्येमे हिमवन्तो महित्वा यस्य समुद्रं रसया सहाहुः ।
यस्येमाः प्रदिशो यस्य बाहू कस्मै देवाय हविषा विधेम ।। ऋ. 10.121.4

yasyeme himvanto mahitvā yasya samudraṁ rasayā sahāhuḥ,

yasyemāḥ pradiśo yasya bāhū kasmai devāya haviṣā vidhema.

[Ime] All these [himavantaḥ] mountains, [samudram] midspheres [saha] along with [rasayā] water bodies [āhuḥ] tell [mahitvā] His greatness; [pradiśaḥ] the directions or dimensions of the space are [yasya] his [bāhū] arms. [vidhema] let us offer [haviṣā] oblation [kasmai devāya] unto Him.

येन द्यौरुग्रा पृथिवी च दृढा येन स्वः स्तभितं येन नाकः ।
यो अन्तरिक्षे रजसो विमानः कस्मै देवाय हविषा विधेम ॥

ऋ. 10.121.5; यजु. 32.6

*yena dyaur ugrā pṛthivī ca dṛḍhā yena svaḥ stabhitaṁ
yena nākaḥ,*
*yo antarikṣe rajaso vimānaḥ kasmai devāya haviṣā
vidhema.*

[Yena] By Whom [dyau] Avakāsa/Chidākāsa was fixed and [ugrā] the chaotic state of [pṛthivī] Akāsa/ bhūtākāsa or space was [dṛḍhā] stabilized; [yena] by Whom the [svaḥ] midspace and interstellar space [stabhitam] was fixed and [nākaḥ] the path of Mokṣa for living beings was made; Who creates [rajasaḥ] all material bodies [antarikṣe] in the space and moves them [vimānaḥ] like flying birds; [vidhema] let us offer [haviṣā] oblation [kasmai] unto that [devāya] Deva from the core of our hearts.

यं क्रन्दसी अवसा तस्तभाने अभ्यैक्षेतां मनसा रेजमाने ।
यत्राधि सूर उदितो विभाति कस्मै देवाय हविषा विधेम ।। ऋ. 10.121.6

*yaṁ krandasī avasā tastabhāne abhyaikṣetaṁ manasā
rejamāne,*
*yatrādhi sūra udito vibhāti kasmai devāya haviṣā
vidhema.*

[Tastabhāne] After the chaotic state of the bhūtākāsa/Akāsa was normalized, Prajāpati [abhyaikṣetāṁ] visualized [manasā] in his mind, [rejamāne] origin of shining [krandasī] stars and planets [avasā] for the sustenance of the biological life in this universe. [Yatrādhi] On the basis of the [yaṁ] same visualization, [sūra] the sun [vibhāti] shines forth [uditaḥ] after its origin. [Vidhema] let us offer [haviṣā] oblations [kasmai devāya] unto Him.

आपो ह यद् बृहतीर्विश्वमायन्गर्भं दधाना जनयन्तीरग्निम् ।
ततो देवानां समवर्ततासुरेकः कस्मै देवाय हविषा विधेम ।। ऋ. 10.121.7

āpo ha yad bṛhatīrviśvamāyangarbhaṁ dadhāna

janayantīrgnim,

tato devānāṁ samavartatāsurekaḥ kasmai devāya
haviṣā vidhema.

[Bṛhatīḥ āpaḥ] The vast prakṛti [homogeneous state of sattva, rajas and tamas] or inactive energy [ekaḥ] lonely [dadhānā] sustained the [āyan viśvam] the whole material world [garbham] in its womb. [Yat] When inactive energy [janyantiḥ] manifested [agni] in active form, [tath] then [asuḥ] origin [devānām] of particles [samvartat] took place. Let us [vidhema] offer [haviṣā] oblation [kasmai devāya] unto Him.

यश्चिदापो महिना पर्यपश्यद्दक्षं दधाना जनयन्तीर्यज्ञम् ।
यो देवेष्वधि देव एक आसीत्कस्मै देवाय हविषा विधेम ।। ऋ. 10.121.8
yaśchidāpo mahinā paryapaśyaddakṣm dadhānā
janayantīryajñam,
yo deveṣvadhi deva eka āsītkasmai devāya haviṣā vidhema.

[Yaḥ] The Prajāpati [chit] when [mahinā] by his power [apaśyat] found [āpaḥ] the undifferentiated prakṛti (inactive energy) [pari] evenly distributed in the chidākāśa [dadhānā] containing [dakṣam] power of creating the universe, He [janyantiḥ] gave rise [yajñam] to the process of creation by activating the energy. [Yaḥ] He [ekaḥ] alone was [devaḥ] the God [deveṣu] of the gods. Let us [vidhema] offer [haviṣā] oblation [kasmai devāya] unto Him.

मा नो हिंसीज्जनिता यः पृथिव्या यो वा दिवं सत्यधर्मा जजान ।
यश्चापश्चन्द्रा बृहतीर्जजान कस्मै देवाय हविषा विधेम ।। ऋ. 10.121.9

mā no hinsījjanitā yaḥ pṛthivyā yo vā divaṁ
satyadharmā jajāna,
yaśchāpaśchandrā bṛhatīrjajāna kasmai devāya haviṣā
vidhema.

May He [mā] not [hinsīta] harm us, [yaḥ] Who [jajān] created [divam] the Chidākāśa; [yaḥ] Who is the [janitā] creator [pṛthivyāḥ] of bhūtākāśa; [satya-dharmā] Who governs the laws of creation; Who [jajāna] induces motion [bṛhatīḥ] into the vast [āpaḥ] inactive energy or undifferentiated prakṛti [chandrā] containing the information of this beautiful [apaḥ] material world. Let us [vidhema] offer [haviṣā] oblation [kasmai devāya] unto Him.

ओं प्रजापते न त्वदेतान्यन्यो विश्वा जातानि परि ता बभूव ।
यत्कामास्ते जुहुमस्तन्नो अस्तु वयं स्याम पतयो रयीणाम् ॥ ऋ .10.121.10

Oṁ prajāpate na tvadetānyanyo viśvā jātāni pari tā bahūva,
yat kāmāste juhumas tanno astu vayaṁ syāma patayo rayīṇām.

[Prajāpate] O Prajāpati [tvat anyaḥ] no one, but You [babhūva] are situated [pari] beyond [etāni] these [viśvā] all [jātāni] originated living and non-living worlds. [Yatkāmāḥ] With whatever desires [juhumaḥ] we offer oblations [te] to you, [tat naḥ astu] let us fulfill them. [Vayam] May we, by your grace, [syāma] be [patayaḥ] the masters [rayīṇām] of all earthly riches and spiritual powers.

Yama Yamī Sūkta

(Ṛgveda 10.10)

Ṛṣi: Mantra No. 1,3,5,7,11,13 Yamī of Vivasvān

        Mantra No. 2,4,8-10, 12,14 Yama of Vivasvān

Devata: Mantra No. 2,4,8-10,12,14 Yamī

        Mantra No. 1,3,5,7,11,13 Yama

Chhanda: Triṣṭup

Personification of nature or cosmos has been the tendency of Vedic seers and Sanskrit poets, right from Vedic period till the time of Kalidāsa and others. They treated nature and cosmos like human beings. The concept of Brahmāṇḍa Puruṣa in Chhāndogya Upaniṣad is a strong proof of it, where the seer has compared the universe with the human body. Yama Yamī Sūkta of the Ṛgveda, one of the Dialogue Sūktas of the Vedas, is also one of such examples.

In this Sūkta, Vivasvān is the vast expanse of universe. The association of particle of matter and antimatter has been depicted in a dialogue form, equating them to male (brother) and female (sister) forms of human beings respectively.

The present Sūkta deals with the twin particles called yama and yamī. Particles herein after called as yama is represented as brother, and its opposite particle herein after called as yamī is represented as his twin sister. This beautiful dialogue, which is cited as prohibiting the union of siblings, is not about the illegitimate relationship at all. This is a dialogue between a particle and its anti-particle. If they join together, they will annihilate each other producing radiation and the material universe will not evolve or expand at all. The Vedas are the knowledge of the existing universe, and sages visualize the important physical processes leading to the present state of universe. The stress here is that the universal creation has taken place due to one particle combining with another particle to generate a third particle, and so on. This is why yamī (sister particle or anti-particle) is advised to join with someone else and not her own opposite particle. The intended sense of mantras can be read as under:

ओ चित्सखांयं सुख्या वंवृत्यां तिरः पुरू चिंदर्णवं जंगन्वान् ।

पितुर्नपांतमा दंधीत वेधा अधि क्षमिं प्रतरं दीध्यांनः ॥ 1 ॥

ō chitsakhāyaṁ sakhyā vavṛtyāṁ tiraḥ purū chidarṇavaṁ jaganvān,
piturnapātamā dadhīta vedhā adhi kṣami prataraṁ dīdhyānaḥ.

1. Yamī (anti-matter particle) speaks: Yamī [anti-particle] having [jaganvān] reached this [arṇavam] space that is [tiraḥ] yet invisible or in darkness [puru chit] and very vast, [vavṛtyām] wanted to join with [sakhāyam] its brother particle born of the same womb, [pituḥ] for making particle the father of [napāt] third generating particle [prataram] endowed with the good qualities [adhi kṣami] in the material world. May [vedhāḥ] God [ā dadhīta] give us power to [didhyānaḥ] generate a third particle for the evolution of universe.

न ते सखां सुख्यं वंष्ट्येतत्सलंक्ष्मा यद्विषुरूपा भवांति ।

महस्पुत्रासो असुरस्य वीरा दिवो धर्तारं उर्विया परिं ख्यन् ॥ 2 ॥

na te sakhā sakhyam vaṣṭyetat salakṣamā yadviṣurūpā bhavāti,
mahasputrāso asurasya vīrā divo dhartāra urviyā pari

khyan.

2. Yama (matter particle) speaks: [te sakhā] Your friend [vaṣṭi] desires [na] not [sakhyam] this friendship, [yat etat] for although [salakṣmā] of one origin (anti-particle is born together from the same womb of energy), [viṣurūpā bhavāti] you are of a different nature (my opposite or anti-particle). [Urviyā mahaḥ divaḥ] This vast space is full of [putrāsaḥ vīrāḥ] other particles [asurasya] of matter [dhartāraḥ] who are the upholders of this space and [pari khyan] enjoy excess over matter particles.

उशन्तिं घा॒ ते अ॒मृता॑स ए॒तदेक॑स्य चि॒त्त्यज॑सं॒ मर्त्य॑स्य ।

नि ते॒ मनो॑ मन॑सि धा॒य्य॒स्मे जन्युः॑ प॒तिस्त॒न्व१ँ॑ मा वि॑विश्याः ॥ 3 ॥

uśanti ghā te amṛtāsa ētadekasya chittyajasaṁ martyasya,
ni te mano manasi dhāyyasme janyuḥ patistanvalmā viviśyāḥ.

3. Yamī speaks: [chit] Though [etad] this friendship is [tyajasaṁ] forbidden [ekasya] only [martyasya] among mortal human beings, the [amṛtāsaḥ] immortals [uśanti] can go desire for [te ghā] this kind of friendship; [te manaḥ] let your mind then [ni dhāyya] concur [asme manasi] with mine, and as [janyuḥ] the progenitor (of material creation) [ā viciṣyāḥ] you enter friendship with [tanvam] me [patiḥ] as husband.

Here the mantra explicitly points out that friendship between mortals born together from the same womb is not allowed, but in the matter of immortals (nature particles) it can take place. So, there is no question of sexual relationship between so called yama and yami as human twins.

न यत्पु॒रा च॑कृ॒मा कद्ध॑ नू॒नमृ॒ता वद॑न्तो॒ अनृ॑तं रपेम ।

गन्धर्वो अप्स्वप्यां च योषा सा नो नाभिः परमं जामि तन्नौ ॥ 4 ॥

na yatpurā chakṛmā kaddha nūnamṛtā vadanto anṛtaṁ rapema,
gandharvo apsvapyā cha yoṣā sā no nābhiḥ paramaṁ jāmi tannau.

4. Yama speaks: We have [na] not [chakṛm] done this [purā] before; We [vadantaḥ] speak [ṛtā] truth and [nūnam] never [rapema] tell [anṛtam] lies. [tat nau] How can we, [jāmi] who are for creation, think of annihilation and become [gandharvaḥ] radiation? [apyā yoṣā] Energy of space originating [apsu] in the space of Brahman is our [nābhiḥ] progenitor.

The mantra says that never a particle born together with its anti-particle joins together for creation to take place. Matter particles in space have their origin in energy particles.

गर्भे नु नौ जनिता दम्पंती कर्देवस्त्वष्टां सविता विश्वरूपः ।

नकिरस्य प्र मिनन्ति व्रतानि वेद नावस्य पृथिवी उत द्यौः ॥5॥

garbhe nu nau janitā dampatī kardevastvaṣṭā savitā viśvarūpaḥ,
nakirasya pra minanti vratāni veda nāvasya pṛthivī uta dyauḥ.

5. Yamī speaks: In the cosmic process of creation, [Devaḥ] God [tvaṣṭā], the creator of universe, [savitā] who inspires us to perform various actions; who is [viṣvarūpaḥ] all-pervading and [janitā] and efficient cause of all [kaḥ] made [nau] us [dampatī] couple [garbhe nu] in the womb of energy itself. [nakī] No one can [pra minanti] break [asya] his [vratāni] laws: [pṛthivī] the earth and [dyau] sun [veda] know [nau] our [asya] this relationship.

को अस्य वेद प्रथमस्याह्नः क ईं ददर्श क इह प्र वोचत् ।

बृहन्मित्रस्य वरुणस्य धाम कदु ब्रव आहनो वीच्या नॄन् ॥ 6 ॥

*ko asya veda prathamasyāhnaḥ ka īm dadarśa ka iha
pra vochat,*
*bṛhanmitrasya varuṇasya dhāma kadu brava āhano
vīchyā nṝn.*

6. Yama speaks: [Kaḥ] Who [veda] knows [prathamasya ahanḥ] the first day [asya] of this birth of ours? [Kaḥ] Who has [dadarśa] beheld [īm] it? [kaḥ] Who has [iha] here [pravochat] revealed it? [Dhāma] The abode [mitrasya] of Mitra (proton) and [varuṇasya] of Varuṇa (electron) has [bṛhat] wide expanse. [Kadu] What do you [bravaḥ] say about it [Āhanaḥ] O Yami, i.e. antiparticle leading [nṛn] particles [vīchyā] to annihilation?

Note: Varuṇa here may be identified as electron and Mitra as proton. Mitra means friend, and protons do sit together like friends inside the nucleus of an atom. The depiction of Mitra as day in the Taittiriya Saṅhitā (2.1.7) also supports this identification. Day stands for brightness and white colour, which was chosen by Vedic sages to represent positive electric charge. Proton carries unit positive charge.

यमस्यं मा यम्यंश् काम् आगन्समाने योनौ सहशेय्याय ।

जायेव पत्यै तन्वं रिरिच्यां वि चिद्वृहेव रथ्येव चक्रा ॥ 7 ॥

*yamasya mā yamyaṁ1 kāma āgansamāne yonau
sahaśeyyāya;*
*jāyeva patye tanvaṁ ririchyāṁ vi chidvṛheva rathyeva
chakrā.*

7. Yamī speaks: [Kāma] Let you yama yearn for living with [mā] me [yamyam] yami. We are [āganta] born from [samāne] the same [yonau] womb [sahaṣeyyāya] to live together. I will join with you, [jāyeva] as a wife

[ririchyām] exposes her [tanvam] body to [patye] her husband; let us [vṛheva] exert ourselves [vi chit] in union [iva] like the [chakrā] two wheels of a [rathyā] chariot.

न तिष्ठन्ति न नि मिषन्त्येते देवानां स्पशं इह ये चरन्ति ।

अन्येन मदाहनो याहि तूयं तेन वि वृंह रथ्येव चक्रा ॥ 8 ॥

na tiṣṭhanti na ni miṣantyete devānāṁ spaśa iha ye charanti,

anyena madāhano yāhi tūyaṁ tena vi vṛha rathyeva chakrā.

8. Yama speaks: [Devānāṁ spaṣa] The spies of the devas, i.e. the virtual particles of electrons [charanti] move around [iha] here, [na tiṣṭhanti] they never stop and [na miṣanti] never close their eyes. [yāhi] Associate [tūyaṁ] quickly, [āhanaḥ] O destructors! [anyena] with some of the particles [tena] who are of your nature [mad] than with me (anti-particle), and exert yourselves in union, like the two wheels of a chariot.

Here the mantra says that yamī (particle) proves destructive in association with her anti-particle (yama), so it was advisable for her to associate with some other particle (of her own nature) to generate a third particle.

Note: Electron is treated like an octopus, which continuously sends its probing arms at an unimaginable rate to feel out its neighborhood. Here in this mantra, spies are virtual particles through which an electron knows what is in its neighbourhood.

रात्रीभिरस्मा अहभिर्दशस्येत्सूर्यस्य चक्षुर्मुहुरुन्निमीयात् ।

दिवा पृथिव्या मिथुना सबन्धू यमीर्यमस्यं बिभृयादजामि ॥ 9 ॥

rātrībhirasmā ahabhirdaśasyet sūryasya chakṣurmuhurunmimīyāt,

divā pṛthivyā mithunā sabandhū yamīryamasya bibhṛyādajāmi.

9. Yamī speaks: [Asmai] To him [yama] let every worshipper [daśasyet] offer oblation both [ahabhiḥ] day and [rātrībhiḥ] night, on him let [chakṣuḥ] the eye of [sūryasya] the sun [muhuḥ] repeatedly [unmimīyāt] rise; (for him may) the [mithunā] kindred pair day and night [sabandhū] unite with [divā] heaven and [pṛthivyāḥ] earth. [Yamī] Yami [bibhrvat] will bear [yamasya] Yama's [ajāmi] unbrotherly (refusal).

आ घा ता गच्छानुत्तरा युगानि यत्रं जामयः कृणवन्नजामि ।

उप बर्बृहि वृषभायं बाहुमन्यमिंच्छस्व सुभगे पतिं मत् ॥ 10 ॥

*ā ghā tā gachchhānuttarā yugāni yatra jāmayaḥ
kṛṇavannajāmi,*
*upa barbṛhi vṛṣabhāya bāhumanyam ichchhasva
subhage patiṁ mat.*

10. Yama speaks: [Uttarā yugāni] The times [ā gacchān] will come, [yatra] when [jāmayaḥ] sister particles [kṛṇvan] will choose [ajāmi] one who is not a brother (anti-particle); [yasmāt] therefore, [subhage] auspicious one, [ichasva] choose [anyam] another particle [mat] except me (anti-particle) [patiṁ bāhum] as husband [upa barbṛhi] to unite with [vṛṣabhāya] for generating a third particle for evolution of universe, nor the union of yami (particle) with its antiparticle [yama] will cause annihilation generating radiation.

किं भ्रातासद्यदनाथं भवाति किमु स्वसा यन्निर्ऋंतिर्निगच्छात् ।

कामंमूता बह्वे३तद्रंपामि तन्वा मे तन्वं३ सं पिंपृग्धि ॥ 11 ॥

*kiṁ bhrātāsadyadanāthaṁ bhavāti kimu svasā
yannirṛtirnigachchhāt,*
*kāmamūtā bahve﹖tadrapāmi tanvā me tanvaṁ? saṁ
pipṛgdhi.*

11. Yamī speaks: [kim asat] What of [bhrātā] brother particle [anti-particle], [yad svasā] whose sister particle

[anātham bhavāti] has no associate? [kimu svasā] What of a sister particle, [yat] who [nigacchāt] cannot alleviate the [nirṛti] problem of her brother particle (antiparticle). [kāmamūtā] Overcome by desire, and [bahu] severally [etat rapāmi] expressing it, I strongly feel [sam piprgdhi] joining [me tanvam] my body [tanvā] with your body.

The Mantra wants to explain, although the union of particles with their anti-particles is destructive, still in the process of creation this type of union takes place usually.

न वा उं ते तन्वां तन्वंश् सं पंपृच्यां पापमांहुर्यः स्वसारं निगच्छांत् ।

अन्येन मत्प्रमुदः कल्पयस्व न ते भ्रातां सुभगे वष्ट्येतत् ॥ 12 ॥

*na vā u te tanvā tanvam1 sam paprchyām
pāpamāhuryaḥ svasāram nigachchhāt,
anyena matpramudaḥ kalpayasva na te bhrātā subhage
vaṣṭyetat.*

12. Yama speaks: [Na] I will not [sam paprchyām] unite [tanvam] myself or my body [te tanvā] with you or your body, [nigachhāt] approaching a [svasāram] sister particle is [āhuḥ] said to be [pāpam] a sin leading to destruction of universe. [Subhage] O auspicious one! [parmudaḥ kalpayasva] enjoy the union of [anyena] someone other [mat] than me; [te bhrātā] your brother particle [na] has no [etat] such [vaṣṭi] desire.

बुतो बंतासि यम नैव ते मनो हृदयं चाविदाम ।

अन्या किल त्वां कक्ष्येव युक्तं परिं ष्वजाते लिबुजेव वृक्षम् ॥13॥

*bato batāsi yama naiva te mano hṛdayam chāvidāma;
anyā kila tvām kakṣayeva yuktam pari ṣvajāte libujeva
vṛkṣam.*

13. Yamī speaks: [Bat] Alas, [asi] you are [bataḥ] weak. [Avidām naiva] I could not fathom [te] your

[manaḥ] mind [cha] or [hṛdayam] your heart. [Anyā kila] Some other particle [pariṣvajāte] embraces [tvām] you [iva] as [libuja] a creeper [yuktam] embraces [vṛkṣam] a tree [kakṣyā] with its thin stems.

Meaning thereby is that an anti-particle cannot remain immune of the union of some particle.

अन्यमू षु त्वं यंम्यन्य उ त्वां परि ष्वजाते लिबुंजेव वृक्षम् ।

तस्यं वा त्वं मनं इच्छा स वा तवाधां कृणुष्व संविदं सुभंद्राम् ॥ 14 ॥

anyamū ṣu tvaṁ yamyanya u tvāṁ pari ṣvajāte
libujeva vṛkṣam,
tasya vā tvaṁ mana ichchhā sa vā tavādhā kṛṇuṣva
saṁvidaṁ subhadrām.

14. Yama speaks: [Yamī] Do you, Yamī, [su] embrace properly [anayam] another; and [anya u] let another [pariṣvājāte] embrace you [libujeva] as a creeper [vṛkṣam] a tree: [tvam ichchha] seek [tasya] his [mana] desire, [sa] let him [vā] also seek [tava] yours; and [kṛṇuṣva] make a [subhadram] happy [saṁvidam] union [adha] leading to the generation of third particle.

Aghamarṣaṇa Sūkta
Ṛgveda (10.190)

ऋतं च सत्यं चाभिद्धात्तपसो·ध्यजायत ।
ततो रात्र्यजायत ततः समुद्रो अर्णवः ॥1॥

ṛtaṁ cha satyaṁ chābhiddhāttapaso·dhyajāyata;
tato rātrayajāyata tataḥ samudro arṇavaḥ.

[Abhi iddhāta] Due to the force [tapasaḥ] of tapas of saṅkalpa, [ṛtam] Ṛta (law of dynamism) [cha] and [satyam] Prakṛti or inactive energy [adhi ajāyata] became active [tataḥ] followed by [rātri] timeless time or incalculable time that is yet to manifest into calculable time [samudro arṇavaḥ] and Bhūtākāśa (space).

Note: In the above mantra, it is informed that

समुद्रादर्णवादधि संवत्सरो अजायत ।
अहोरात्राणि विदधत् विश्वस्य मिषतो वशी ॥2॥

samudrādarṇavādadhi saṁvatsaro ajāyata;
ahorātrāṇi vidadhat viśvasya miṣato vaśī.

[Samudrad arṇavād adhi] From Bhūtākāśa (dynamic space) as base [saṁvatsaraḥ] calculable time [ajāyat] appeared or manifested. Thus, the [vaśī] Governor [viśvasya] of universe [vidadhat] fixed the time period of [ahaḥ] creation and [rātrāṇi] decreation [miṣataḥ] as the blinking of His eyes.

Here, the technical term 'Ṛta' is formed of the root √ṛ meaning 'gati', i.e. 'to go'. So, 'Ṛta' here represents the 'law of dynamism' and the technical term 'Satya' represents 'Prakṛti' or 'inactive energy'. Rātrī represents incalculable time that is yet to manifest and the technical term 'samudra' in the Veda, according to Vedic scholar Yāska, means space. Arṇava again is formed of the root √ṛ meaning 'gati', i.e. 'to go'. 'Arṇava' has been used as the attributive of samudra. So, 'samudro arṇavaḥ' means 'Bhūtākāśa or space with speed' or dynamic space or expanding space between stars and planets.

Thus, according to the Vedas, Ṛta is the principle of dynamism that applies when spontaneous agitation of prakṛti or inactive energy takes place. This dynamism exists in the very nature of Brahman. He does not want to be static, rather as dynamic. He wants the change of state. This want of 'change of state' by Brahman has been expressed in several Vedic statements like एकोऽहं बहु स्याम - eko'ham bahu syām, etc. This desire for change comes spontaneously in Brahman without any condition, situational, or subjective whatsoever. Nature and we all have inherited this habit of change from the Brahman Himself. So natural law or will of God is one and the same thing. It is called natural law from an objective perspective, but the same is called the will of Brahman (God) from a spiritual or subjective angle or consideration. Natural law is also known as the law of divine or divine law. Coming back to the same point, it can be observed that in the state of prak,ti, all-natural forces were also unified into one fundamental force. However, during the state of vikṛti due to the principle of Ṛta, the grand unification of forces also disintegrates and they start separating and manifest into various forms undergoing to various phases of change till the atoms are formed and come together to make molecules, molecules collecting to create cells, and groups of cells forming tissues and organs and organisms.

Now the question arises as to what makes atoms,

molecules, and cells come together as they do? The answer is the principle of 'Rta' or say 'law of change of nature' or 'law of dynamism'. This 'law of dynamism' may also be called as the principle of māyā. The change of state takes place by permutations and combinations of various components. Although at the different change of phases, there seem to be different laws at work, but they all express the same unified 'law of Rta or Māyā' which permeates them all and which remains unchanged, which is the source of all forces, energy and particles and a field of pure existence. This unified law contains the details of all natural laws in the seed form. Theoretician physicists and mathematicians try to describe them through their formulae and equations. The universe thus is the cosmic shape of Prakṛti (inactive energy). Just as a seed contains the information of a tree in pure potentiality; the tree is the manifest shape of inner hidden dynamics available in the seed, similarly, Prakṛti (energy) contains the information of universe in pure potentiality, and the universe is the manifest shape of inner dynamics (sattva/intelligence, rajas/motion, and tamas/inertia) of prakṛti or inacative energy. Thus, prak,ti is the seed of the universe which grows into a universe following principle of Rta from the creator with goals and intentions. So, at one level (unified level/unmanifest level), natural law is pure prakṛti (equilibrium state of sattva or intelligence, raja or motion, and tamas or inertia or inaction; on another (diversified level/manifest level) it is particles and matter; then it is unicellular and multicellular organisms.

सूर्य्याचन्द्रमसौ धाता यथापूर्वमकल्पयत् ।
दिवं च पृथिवीं चान्तरिक्षमथो स्वः ॥ ऋ .10.190.3

sūryyācandramasau dhātā yathāpūrvam akalpayat.
divaṁ ca pṛthivīṁ cāntarikṣam atho svaḥ.

[Dhāta] Brahman [akalpayat] created [divam] Bhūtākāśa (dynamic space) [sūryāchandramasau] the sun, the moon, [pṛthivīm] the earth, [antarikṣam] the magnetosphere of the earth [atha] and [svaḥ] magnetic field of the sun in our solar system [yathāpūrvam] in the same manner as previous solar systems.

Indra-Sūrya Sūkta

Ṛgveda (5.40.5-9)

Mantras dealing with Total Solar Eclipse

यत्त्वा सूर्य स्वर्भानुस्तमसाविध्यदासुरः।

अक्षेत्रविद्यथा मुग्धो भुवनान्यदीधयुः॥ 5 ॥

yattvā sūrya *svarbhānustamasāvidhyadāsuraḥ;*
akṣetravidyathā mugdho bhuvanānyadīdhayuḥ.

The above mantra is talking about the total solar eclipse.

[Yat tvā] When [sūrya] the sun was [avidhyat] covered/overlapped by [tamasā] the darkness [āsuraḥ svarbhānu] of new moon, [bhuvanāni] the people in the world [adīdhayuḥ] become [mugdhaḥ] stranded [akṣetravid] and do not recognize places.

The above mantra deals with total solar eclipse.

स्वर्भानोरध यदिन्द्र माया अवो दिवो वर्तमाना अवाहन्।

गूळ्हं सूर्यं तमसापव्रतेन तुरीयेण ब्रह्मणाविन्ददत्रिः॥ 6 ॥

svarbhānoradha yadindra māyā avo divo vartamānā
avāhan,
gūḷhaṃ sūryaṃ tamasāpavratena turīyeṇa
brahmaṇāvindadatriḥ.

[Yat] When [Indra] the sun was [avavāhan] covered by [svarbhānu] new moon [adha] from below with her [māyā] umbra (the darkest part of the shadow of the moon) [divaḥ] in the celestial sphere, [atriḥ apavraten] the descending node of Moon [avindat] discovered the sun [tamasā gūḍham] concealed by the umbra [turīyeṇa brahmaṇā] in the fourth stage. According to the Ṛgveda, there are four stages of total solar eclipse:

1. The Moon starts covering/overlapping the sun.

2. The entire disk of the Sun is covered by the Moon, Only the Sun's corona is visible. Giving the diamond ring effect.

3. The Moon starts moving away, and the Sun reappears.

4. The Moon stops overlapping the Sun's disk. The eclipse ends at this stage.

मा मामिमं तव सन्तमत्र इरस्या द्रुग्धो भियसा नि गारीत् ।
लं मित्रो असि सत्यराधास्तौ मेहावतं वरुणश्च राजा ॥ 7 ॥

*mā māmimaṁ tava santamatra irasyā drugdho bhiyasā
ni gārīt,*

*tvaṁ mitro asi satyarādhāstau mehāvataṁ varuṇaśca
rājā.*

[Mā] Let not [atre] the Moon [irasyā] have desire to [nigārīt] swallow [mām] me with her [bhiyasā] fearful (darkness), as [tava santam] I am in your shelter, [tvam] you are [mitraḥ] my friend who is on the true path (around the earth). [Tau] Do you [cha] and [rājā varuṇa] the royal Varuṇa (the Sun) both [avatam] protect [mā] me [iha] here on this globe.

Here in view of the celestial process of total solar eclipse accomplished by the New Moon and the Sun, it is prayed that may both the Moon and the Sun protect the whole creation on the earth which is under their shelter.

ग्राव्णो ब्रह्मा युयुजानः सपर्यन्कीरिणा देवान्नमसोपशिक्षन् ।
अत्रिः सूर्यस्य दिवि चक्षुराधात्स्वर्भानोरप माया अघुक्षत् ॥8॥

grāvṇo brahmā yuyujānaḥ saparyankīriṇā

devānnamasopaśikṣan,
atriḥ sūryasya divi cakṣurādhātsvarbhānorapa māyā
aghukṣat.

Again it has been suggested that [Brahma Atriḥ] A seeker or a Vedic scholar [yuyujānaḥ] who wants to attain knowledge; who has [saparyan] appreciation [kīriṇā] for knowledge and [upaśikṣan] desirous of learning the [devān] astronomical science [namasā] with humility, [chakṣur ādadhīt] should place his focus [grāvaṇaḥ] on the illumination [sūryasya] of the Sun [divi] in the sky, all his [māyā] delusions [svarbhānoḥ] regarding celestial process of solar eclipse [apa aghukṣat] will be dispersed.

Note: In Yogadarśana also it is stated: सूर्ये संयमात् भुवन ज्ञानम् । When a yogī do sañyama in the Sun, he is able to know the formation of entire universe.

यं वै सूर्यं स्वर्भानुस्तमसाविध्यदासुरः ।
अत्रयस्तमन्वविन्दन्नह्यन्ये अशक्नुवन् ॥ 9 ॥
yam vai sūryam svarbhānustamasāvidhyadāsuraḥ,
atrayastamanvavindannahyanye aśaknuvan.

[Yam] The celestial phenomenon of solar eclipse where [sūryam] sun is [avidhyat] concealed by [svarbhānū] the Moon [tamasā] with her umbra (darkness), is [avindan] understood by [atrayaḥ] the scholars who are well versed in astronomy, [anye] other lay persons are [aśaknuvan] not able to understand this secret of nature.